United States Government Accountability Office

Testimony
Before the Subcommittee on Government Operations
Committee on Oversight and Government Reform
House of Representatives

For Release on Delivery
Expected at 2:30 p.m. EDT
Tuesday, May 14, 2013

DATA CENTER CONSOLIDATION

Strengthened Oversight Needed to Achieve Billions of Dollars in Savings

Statement of David A. Powner, Director
Information Technology Management Issues

GAO-13-627T

GAO Highlights

Highlights of GAO-13-627T, a testimony before the Subcommittee on Government Operations, Committee on Oversight and Government Reform, House of Representatives

May 2013

DATA CENTER CONSOLIDATION

Strengthened Oversight Needed to Achieve Billions of Dollars in Savings

Why GAO Did This Study

In 2010, as focal point for information technology management across the government, OMB's Federal Chief Information Officer launched the Federal Data Center Consolidation Initiative—an effort to consolidate the growing number of federal data centers. In July 2011 and July 2012, GAO evaluated 24 agencies' progress and reported that nearly all of the agencies had not completed a data center inventory or consolidation plan and recommended that they do so.

GAO was asked to testify on its report, being released today, that evaluated agencies' reported progress against OMB's planned consolidation and cost savings goals, and assessed the extent to which the oversight organizations put in place by OMB for the Federal Data Center Consolidation Initiative are adequately performing oversight of agencies' efforts to meet these goals. In this report, GAO assessed agencies' progress against OMB's goals, analyzed the execution of oversight roles and responsibilities, and interviewed OMB, GSA, and Data Center Consolidation Task Force officials about their efforts to oversee agencies' consolidation efforts.

What GAO Recommends

In its report, GAO recommended that OMB's Federal Chief Information Officer track and report on key performance measures, extend the time frame for achieving planned cost savings, and improve the execution of important oversight responsibilities. OMB agreed with two of GAO's recommendations and plans to evaluate the remaining recommendation related to extending the time frame.

View GAO-13-627T. For more information, contact David A. Powner at (202) 512-9286 or pownerd@gao.gov.

What GAO Found

The 24 agencies participating in the Federal Data Center Consolidation Initiative made progress towards the Office of Management and Budget's (OMB) goal to close 40 percent, or 1,253 of the 3,133 total federal data centers, by the end of 2015, but OMB has not measured agencies' progress against its other goal of $3 billion in cost savings by the end of 2015. Agencies closed 420 data centers by the end of December 2012, and have plans to close an additional 548 to reach 968 by December 2015—285 closures short of OMB's goal. OMB has not determined agencies' progress against its cost savings goal because, according to OMB staff, the agency has not determined a consistent and repeatable method for tracking cost savings. This lack of information makes it uncertain whether the $3 billion in savings is achievable by the end of 2015. Until OMB tracks and reports on performance measures such as cost savings, it will be limited in its ability to oversee agencies' progress against key goals.

Pursuant to OMB direction, three organizations—the Data Center Consolidation Task Force, the General Services Administration (GSA) Program Management Office, and OMB—are responsible for federal data center consolidation oversight activities; while most activities are being performed, there are still several weaknesses in oversight. Specifically,

- While the Data Center Consolidation Task Force has established several initiatives to assist agencies in their consolidation efforts, such as holding monthly meetings to facilitate communication among agencies, it has not adequately overseen its peer review process for improving the quality of agencies' consolidation plans.
- The GSA Program Management Office has collected agencies' quarterly data center closure updates and made the information publically available on an electronic dashboard for tracking consolidation progress, but it has not fully performed other oversight activities, such as conducting analyses of agencies' inventories and plans.
- OMB has implemented several initiatives to track agencies' consolidation progress, such as establishing requirements for agencies to update their plans and inventories yearly and to report quarterly on their consolidation progress. However, the agency has not approved the plans on the basis of their completeness or reported on progress against its goal of $3 billion in cost savings.

The weaknesses in oversight of the data center consolidation initiative are due, in part, to OMB not ensuring that assigned responsibilities are being executed. Improved oversight could better position OMB to assess progress against its cost savings goal and minimize agencies' risk of not realizing expected cost savings.

In March 2013, OMB issued a memorandum that integrated the Federal Data Center Consolidation Initiative with the PortfolioStat initiative, which requires agencies to conduct annual reviews of its information technology investments and make decisions on eliminating duplication, among other things. The memorandum also made significant changes to the federal data center consolidation effort, including the initiative's reporting requirements and goals. Specifically, agencies are no longer required to submit the previously required consolidation plans and the memorandum does not identify a cost savings goal.

United States Government Accountability Office

Chairman Mica, Ranking Member Connolly, and Members of the Subcommittee:

I am pleased to be here today to discuss federal agencies' continuing efforts to consolidate their data centers. As federal agencies have modernized their information technology (IT) operations, put more of their services online, and increased their information security profiles, their need for computing power and data storage resources has resulted in a dramatic increase in the federal data centers and a corresponding increase in operational costs. In response, the Office of Management and Budget's (OMB) Federal Chief Information Officer (CIO) launched the Federal Data Center Consolidation Initiative (FDCCI) in 2010.

Over the past few years, we have reported and testified[1] on federal data center consolidation, noting that, while the initiative has the potential to provide billions of dollars in savings, improvements in the oversight of agencies' efforts are needed. For example, in July 2012, we reported on the progress the 24 participating federal departments and agencies[2] were making on this initiative and found that, while progress had been made, nearly all of the agencies had not yet completed a data center inventory or the consolidation plans needed to implement their consolidation initiatives.

---

[1] GAO, *Information Technology: OMB and Agencies Need to Fully Implement Major Initiatives to Save Billions of Dollars*, GAO-13-297T (Washington, D.C.: Jan. 22, 2013); *Data Center Consolidation: Agencies Making Progress on Efforts, but Inventories and Plans Need to Be Completed*, GAO-12-742 (Washington, D.C.: July 19, 2012); *Information Technology Reform: Progress Made; More Needs to Be Done to Complete Actions and Measure Results*, GAO-12-745T (Washington, D.C.: May 24, 2012); *Information Technology Reform: Progress Made; More Needs to Be Done to Complete Actions and Measure Results*, GAO-12-461 (Washington, D.C.: Apr. 26, 2012); and *Data Center Consolidation: Agencies Need to Complete Inventories and Plans to Achieve Expected Savings*, GAO-11-565 (Washington, D.C.: July 19, 2011).

[2] The 24 major departments and agencies that participate in the Federal Data Center Consolidation Initiative are the Departments of Agriculture, Commerce, Defense, Education, Energy, Health and Human Services, Homeland Security, Housing and Urban Development, the Interior, Justice, Labor, State, Transportation, the Treasury, and Veterans Affairs; the Environmental Protection Agency, General Services Administration, National Aeronautics and Space Administration, National Science Foundation, Nuclear Regulatory Commission, Office of Personnel Management, Small Business Administration, Social Security Administration, and U.S. Agency for International Development.

You asked us to testify on our report being released today that evaluated agencies' reported progress against OMB's planned consolidation and cost savings goals and assessed the extent to which the oversight organizations put in place by OMB for FDCCI are adequately performing oversight of agencies' efforts to meet these goals.[3] This report contains a detailed overview of our scope and methodology, including the steps we took to assess the quality of data that we relied on.

All work on which this testimony is based was performed in accordance with generally accepted government auditing standards. Those standards require that we plan and perform the audit to obtain sufficient, appropriate evidence to provide a reasonable basis for our findings and conclusions based on our audit objectives. We believe that the evidence obtained provides a reasonable basis for our findings and conclusions based on our audit objectives.

Background

Over the last 15 years, the federal government's increasing demand for IT has led to a dramatic rise in the number of federal data centers and a corresponding increase in operational costs. According to OMB, the federal government had 432 data centers in 1998 and more than 1,100 in 2009. Operating such a large number of centers is a significant cost to the federal government, including costs for hardware, software, real estate, and cooling. For example, according to the Environmental Protection Agency, the electricity cost to operate federal servers and data centers across the government is about $450 million annually. According to the Department of Energy, data center spaces can consume 100 to 200 times more electricity than a standard office space. According to OMB, reported server utilization rates as low as 5 percent and limited reuse of data centers within or across agencies lend further credence to the need to restructure federal data center operations to improve efficiency and reduce costs.

---

[3]GAO, *Data Center Consolidation: Strengthened Oversight Needed to Achieve Cost Savings Goal*, GAO-13-378 (Washington, D.C.: Apr. 23, 2013).

OMB and the Federal CIO Established the Federal Data Center Consolidation Initiative

Concerned about the size of the federal data center inventory and the potential to improve the efficiency, performance, and the environmental footprint of federal data center activities, OMB, under the direction of the Federal CIO, established FDCCI in February 2010. This initiative's four high-level goals are to

- promote the use of "green IT"[4] by reducing the overall energy and real estate footprint of government data centers;

- reduce the cost of data center hardware, software, and operations;

- increase the overall IT security posture of the government; and

- shift IT investments to more efficient computing platforms and technologies.

As part of FDCCI, OMB required the 24 agencies to identify a senior, dedicated data center consolidation program manager to lead their agency's consolidation efforts. In addition, agencies were required to submit an asset inventory baseline and other documents that would result in a plan for consolidating their data centers. The asset inventory baseline was to contain detailed information on each data center and identify the consolidation approach to be taken for each one. It would serve as the foundation for developing the final data center consolidation plan. The data center consolidation plan would serve as a technical road map and approach for achieving the targets for infrastructure utilization, energy efficiency, and cost efficiency.

While OMB is primarily responsible for FDCCI, the agency designated two agency CIOs to be executive sponsors to lead the effort within the Federal CIO Council,[5] the principal interagency forum to improve IT-related practices across the federal government. In addition, OMB identified two additional organizations to assist in managing and overseeing FDCCI:

---

[4]"Green IT" refers to environmentally sound computing practices that can include a variety of efforts, such as using energy efficient data centers, purchasing computers that meet certain environmental standards, and recycling obsolete electronics.

[5]As of February 2013, OMB had assigned one CIO from the Department of the Interior. Initially there had been two executive sponsors, but one resigned and OMB stated that they had no plans to fill the second position.

- The GSA FDCCI Program Management Office is to support OMB in the planning, execution, management, and communications for FDCCI.

- The Data Center Consolidation Task Force is comprised of the data center consolidation program managers from each agency. According to its charter, the Task Force is critical to supporting collaboration across the FDCCI agencies, including identifying and disseminating key pieces of information, solutions, and processes that will help agencies in their consolidation efforts.

In an effort to accelerate federal data center consolidation, OMB has directed agencies to use cloud computing[6] as an approach to migrating or replacing systems with Internet-based services and resources. In December 2010, in its 25 Point IT Reform Plan,[7] OMB identified cloud computing as having the potential to play a major part in achieving operational efficiencies in the federal government's IT environment. According to OMB, cloud computing brings a wide range of benefits, including that it is (1) economical—a low initial investment is required to begin and additional investment is needed only as system use increases, (2) flexible—computing capacity can be quickly and easily added or subtracted, and (3) fast—long procurements are eliminated, while providing a greater selection of available services. To help achieve these benefits, OMB issued a "Cloud First" policy that required federal agencies to increase their use of cloud computing whenever a secure, reliable, and cost-effective cloud solution exists.

We have previously reported that, while selected federal agencies had made progress in implementing cloud computing, they also faced challenges.[8] For example, agencies identified cloud computing challenges related to meeting federal security requirements that are

---

[6]Cloud computing is an emerging form of delivering computing services via networks with the potential to provide IT services more quickly and at a lower cost. Among other things, it provides users with on-demand access to a shared and scalable pool of computing resources with minimal management effort or service provider interaction.

[7]OMB, *25 Point Implementation Plan to Reform Federal Information Technology Management* (Washington, D.C.: Dec. 9, 2010).

[8]GAO, *Information Technology Reform: Progress Made but Future Cloud Computing Efforts Should be Better Planned*, GAO-12-756 (Washington, D.C.: July 11, 2012) and *Information Security: Federal Guidance Needed to Address Control Issues with Implementing Cloud Computing*, GAO-10-513 (Washington, D.C.: May 27, 2010).

unique to government agencies, such as continuous monitoring and maintaining an inventory of systems. Agencies also noted that, because of the on-demand, scalable nature of cloud services, it can be difficult to define specific quantities and costs and, further, that these uncertainties make contracting and budgeting difficult due to the fluctuating costs associated with scalable and incremental cloud service procurements. Finally, agencies cited other challenges associated with obtaining guidance, and acquiring knowledge and expertise, among other things.

More recently, in March 2013, OMB issued a memorandum documenting the integration of FDCCI with the PortfolioStat initiative.[9] Launched by OMB in March 2012, PortfolioStat requires agencies to conduct an annual agency-wide IT portfolio review to, among other things, reduce commodity IT[10] spending, demonstrate how its IT investments align with the agency's mission and business functions, and make decisions on eliminating duplication.[11] OMB's March 2013 memorandum discusses OMB's efforts to further the PortfolioStat initiative by incorporating several changes, such as consolidating previously collected IT-related plans, reports, and data submissions. The memorandum also establishes new agency reporting requirements and related time frames. Specifically, agencies are no longer required to submit the data center consolidation plans previously required under FDCCI. Rather, agencies are to submit information to OMB via three primary means—an information resources management strategic plan,[12] an enterprise road map,[13] and an integrated data collection channel.[14]

---

[9] OMB, *Fiscal Year 2013 PortfolioStat Guidance: Strengthening Federal IT Portfolio Management*, Memorandum M-13-09 (Washington, D.C.: Mar. 27, 2013).

[10] According to OMB, commodity IT includes services such as IT infrastructure (data centers, networks, desktop computers and mobile devices); enterprise IT systems (e-mail, collaboration tools, identity and access management, security, and web infrastructure); and business systems (finance, human resources, and other administrative functions).

[11] OMB, *Implementing PortfolioStat,* Memorandum M-12-10 (Washington D.C.: Mar. 30, 2012).

[12] OMB, *Management of Federal Information Resources*, Circular A-130 (Washington, D.C.: Nov. 30, 2000). According to OMB Circular A-130, an agency's information resources management strategic plan should describe how information resources management activities help accomplish agency missions, and ensure that information resource management decisions are integrated with organizational planning, budget, procurement, financial management, human resources management, and program decisions.

GAO Has Previously Reported on Significant Weaknesses in Agencies' Inventories and Plans

In July 2012, we issued a report on the status of FDCCI and found that, while agencies' 2011 inventories and plans had improved as compared to their 2010 submissions, significant weaknesses still remained.[15] Specifically, while all 24 agencies reported on their inventories to some extent, only 3 had submitted a complete inventory.[16] The remaining 21 agency submissions had weaknesses in several areas. For example, while most agencies provided complete information on their virtualization[17] efforts, network storage, and physical servers, 18 agencies did not provide complete data center information, such as data center type, gross floor area, and target date for closure. In particular, several agencies fully reported on gross floor area and closure information, but partially reported data center costs. In addition, 17 agencies did not provide full information on their IT facilities and energy usage. For example, the Department of Labor partially reported on total data center IT power capacity and average data center electricity usage and did not report any information on total data center power capacity. We also noted that 3 agencies had submitted their inventory using an outdated format, in part, because OMB had not publicly posted its revised guidance. Figure 1 provides an assessment of the completeness of agencies' 2011 inventories, by key element.

---

[13]OMB, *Increasing Shared Approaches to Information Technology Services* (Washington, D.C.: May 2, 2012). The enterprise road map is to include a business and technology architecture, an IT asset inventory, a commodity IT consolidation plan, a line of business service plan, and an IT shared service plan.

[14]The integrated data collection channel will be used by agencies to report structured information, such as progress in meeting IT strategic goals, objectives, and metrics, as well as cost savings and avoidances resulting from IT management actions.

[15]GAO-12-742.

[16]These agencies were the Department of Housing and Urban Development, the National Science Foundation, and the Social Security Administration.

[17]Virtualization is a technology that allows multiple, software-based machines with different operating systems, to run in isolation, side-by-side, on the same physical machine.

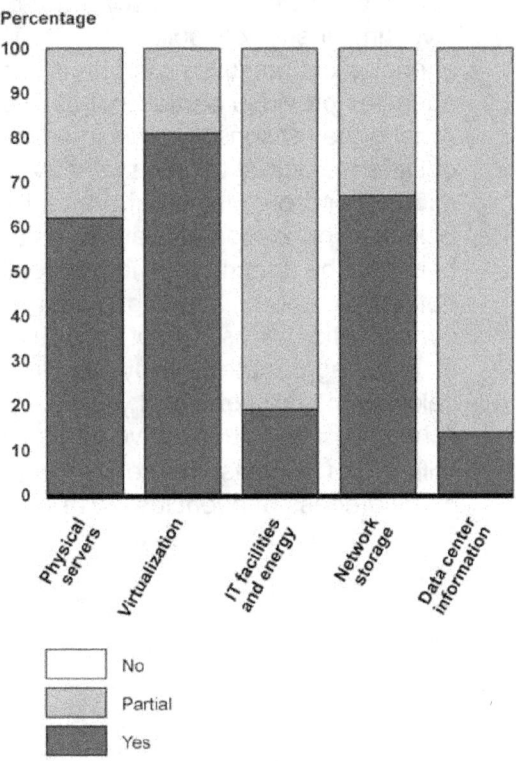

Figure 1: Twenty-one Agencies' Completion of Required Information for Data Center Inventory Key Elements, as of June 2011

Source: GAO analysis of agency data.

Officials from several agencies reported that some of the required information was unavailable at certain data center facilities. We reported that, because the continued progress of FDCCI is largely dependent on accomplishing goals built on the information provided by agency inventories, it will be important for agencies to continue to work on completing their inventories, thus providing a sound basis for their savings and utilization forecasts.

In addition, while all 24 agencies submitted consolidation plans to OMB, only 1 had submitted a complete plan.[18] For the remaining 23 agencies, selected elements were missing from each plan. For example, among the

---

[18]This agency was the Department of Commerce.

24 agencies, all provided complete information on their qualitative impacts, and nearly all included a summary of the consolidation approach, a well-defined scope for data center consolidation, and a high-level timeline for consolidation efforts. However, most notably, 21 agencies did not fully report their expected cost savings; of those, 13 agencies provided partial cost savings information and 8 provided none. Among the reasons that this information was not included, a Department of Defense official told us that it was challenging to gather savings information from all the department's components, while a National Science Foundation official told us the information was not included because the agency had not yet realized any cost savings and so had nothing to report. Other significant weaknesses were that many agencies' consolidation plans did not include a full cost-benefit analysis that included aggregate year-by-year investment and cost savings calculations through fiscal year 2015, a complete master program schedule,[19] and quantitative goals, such as complete savings and utilization forecasts. Figure 2 provides an assessment of the completeness of agencies' 2011 consolidation plans, by key element.

---

[19]A master program schedule was to be created for the entire agency, from the detailed implementation schedules provided by each of the data center managers as well as driven by related federal government activities (e.g., OMB reporting, budget submission, or beginning of a new fiscal year).

Figure 2: Twenty-four Agencies' Completion of Required Information for Data Center Consolidation Plan Key Elements, as of September 2011

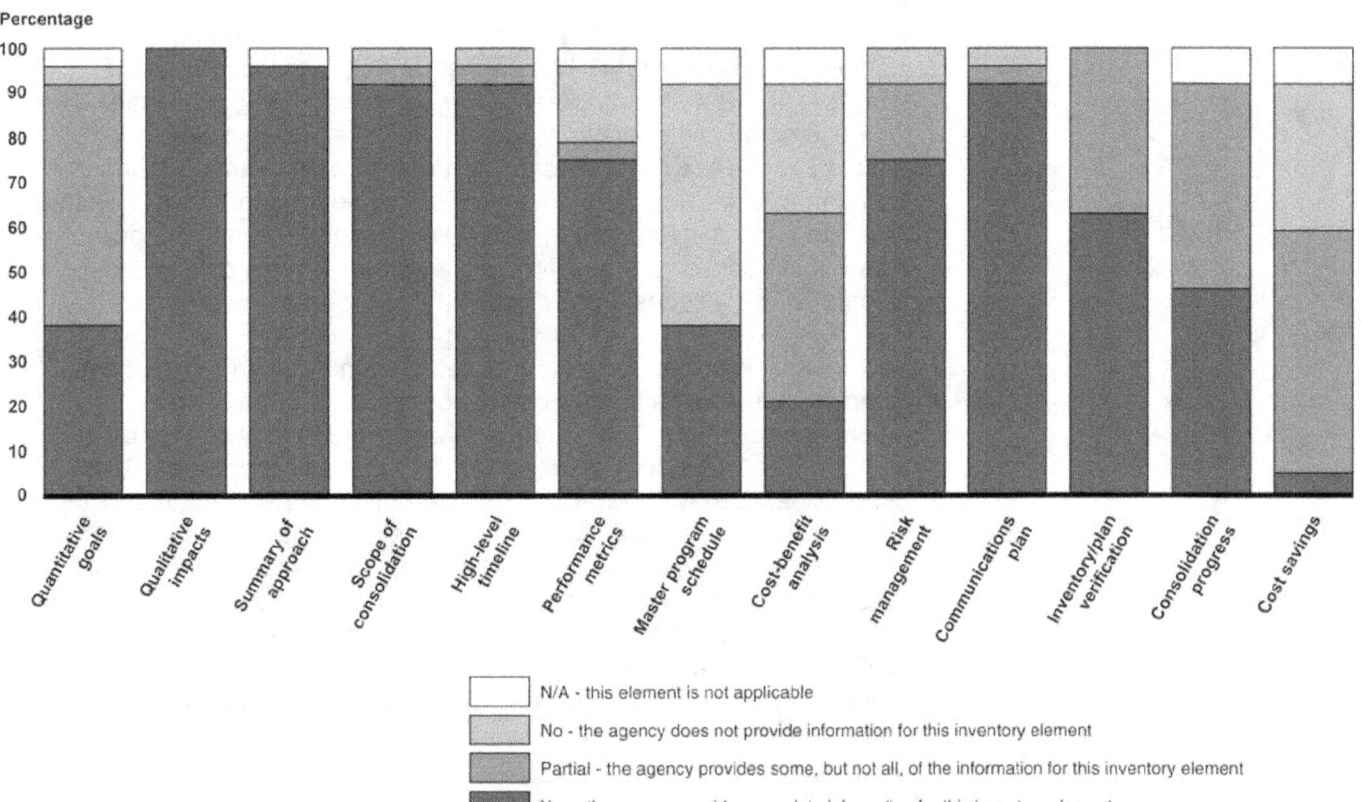

Source: GAO analysis of agency data.

Officials from several agencies reported that the plan information was still being developed. We concluded that, in the continued absence of completed consolidation plans, agencies are at risk of implementing their respective initiatives without a clear understanding of their current state and proposed end state and not being able to realize anticipated savings, improved infrastructure utilization, or energy efficiency.

We also found that while agencies were experiencing data center consolidation successes, they were also encountering challenges. While almost 20 areas of success were reported, the 2 most often cited focused on virtualization and cloud services as consolidation solutions, and working with other agencies and components to find consolidation opportunities. Further, while multiple challenges were reported, the two most common challenges were both specifically related to FDCCI data reporting required by OMB: obtaining power usage information and providing a quality data center asset inventory.

We further reported that, to assist agencies with their data center consolidation efforts, OMB had sponsored the development of a FDCCI total cost of ownership model that was intended to help agencies refine their estimated costs for consolidation; however, agencies were not required to use the cost model as part of their cost estimating efforts. We stated that, until OMB requires agencies to use the model, agencies will likely continue to use a variety of methodologies and assumptions in establishing consolidation estimates, and it will remain difficult to summarize projections across agencies.

Accordingly, we reiterated our prior recommendation that agencies complete missing plan and inventory elements and made new recommendations to OMB to publically post guidance updates on the FDCCI website and to require agencies to use its cost model. OMB generally agreed with our recommendations and has since taken steps to address them. More specifically, OMB posted its 2012 guidance for updating data center inventories and plans, as well as guidance for reporting consolidation progress, to the FDCCI public website. Further, the website has been updated to provide prior guidance documents and OMB memoranda. In addition, OMB's 2012 consolidation plan guidance requires agencies to use the cost model as they develop their 2014 budget request.

Consolidation of Federal Data Centers is Under Way, but Initiativewide Cost Savings Have Not Been Determined

We and other federal agencies[20] have documented the need for initiatives to develop performance measures to gauge progress. According to government and industry leading practices, performance measures should be measurable, outcome-oriented, and actively tracked and reported. For FDCCI, OMB originally established goals for data center closures and the expected cost savings. Specifically, OMB expected to consolidate approximately 40 percent of the total number of agency data centers and achieve $3 billion in cost savings by the end of 2015, and established the means of measuring performance against those goals through several methods.

The 24 agencies have collectively made progress towards OMB's data center consolidation goal to close 40 percent, or approximately 1,253 of the 3,133 data centers, by the end of 2015. To track their progress, OMB requires agencies to report quarterly on their completed and planned performance against that goal via an online portal. After being reviewed for data quality and security concerns, the GSA FDCCI Program Management Office makes the performance information available on the federal website dedicated to providing the public with access to datasets developed by federal agencies, http://data.gov.

As of February 2013, agencies had collectively reported closing a total of 420 data centers by the end of December 2012,[21] and were planning to close an additional 396 data centers—for a total of 816—by September 2013.[22] While the number of data centers that agencies are planning to

---

[20]GAO, *Aviation Weather: Agencies Need to Improve Performance Measurement and Fully Address Key Challenges*, GAO-10-843 (Washington, D.C.: Sept. 9, 2011); GAO, *NextGen Air Transportation System: FAA's Metrics Can Be Used to Report on Status of Individual Programs, but Not of Overall NextGen Implementation or Outcomes*, GAO-10-629 (Washington, D.C.: July 27, 2010); OMB, *Guide to the Program Assessment Rating Tool* (Washington, D.C.: January 2008); Department of the Navy, Office of the Chief Information Officer, *Guide for Developing and Using IT Performance Measurements* (Washington, D.C.: October 2001); and GSA, *Performance-Based Management: Eight Steps To Develop and Use Information Technology Performance Measures Effectively* (Washington, D.C.: 1996).

[21]Of the 24 agencies, 17 reported closing at least one data center by the end of December 2012. The remaining 7 agencies did not report closing any data centers during this time.

[22]Of the 24 agencies, 19 reported plans to close at least one data center by the end of September 2013. The remaining 5 agencies did not report plans to close any data centers in this time frame.

close from October 2013 through December 2015 (the planned completion date of FDCCI) is not reported on http://data.gov, OMB's July 2012 quarterly report to Congress[23] on the status of federal IT reform efforts contains other information on agencies' data center closure plans. Among other things, the report states that agencies have collectively committed to closing a total of 968 data centers by the end of 2015. According to OMB staff from the Office of E-Government and Information Technology, this figure represents the number of commitments reported by agencies, as compared to the initiative's overall goal of closing 1,253 data centers by December 2015. The agencies have not identified the remaining 285 consolidation targets to achieve that goal. OMB's January 2013 quarterly report to Congress[24] does not provide any new information about either planned or completed agency data center closures. See figure 3 for a graphical depiction of agencies' progress against OMB's data center consolidation goal.

---

[23]OMB, *Quarterly Report to Congress: Integrated, Efficient, and Effective Uses of Information Technology* (Washington, D.C.: July 30, 2012).

[24]OMB, *Quarterly Report to Congress: Integrated, Efficient, and Effective Uses of Information Technology* (Washington, D.C.: Jan. 31, 2013).

Figure 3: Agencies' Progress against OMB's Data Center Consolidation Goal (as of February 2013)

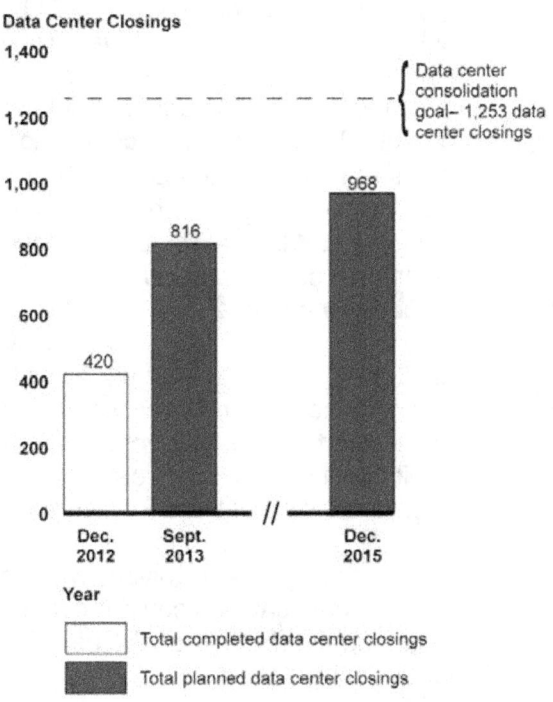

Source: GAO analysis of OMB and GSA data.

However, OMB has not measured agencies' progress against the cost savings goal of $3 billion by the end of 2015. According to a staff member from OMB's Office of E-Government and Information Technology, as of November 2012, the total savings to date had not been tracked but were believed to be minimal. The staff member added that, although data center consolidation involves reductions in costs for existing facilities and operations, it also requires investment in new and upgraded facilities and, as a result, any current savings are often offset by the reinvestment of those funds into ongoing consolidation efforts. Finally, the staff member stated that OMB recognizes the importance of tracking cost savings and is working to identify a consistent and repeatable method for tracking cost savings as part of the integration of FDCCI with PortfolioStat, but stated that there was no time frame for when this would occur.

The lack of initiativewide cost savings data makes it unclear whether agencies will be able to achieve OMB's projected savings of $3 billion by the end of 2015. In previous work, we found that agencies' cost savings projections were incomplete and, in some cases, unreliable. Specifically, in July 2012,[25] we reported that most agencies had not reported their expected cost savings in their 2011 consolidation plans. Officials from several agencies reported that this information was still being developed. Notwithstanding these weaknesses, we found that agencies collectively reported anticipating about $2.4 billion in cumulative cost savings by the end of 2015 (the planned completion date of FDCCI).[26] With less than 3 years remaining to the 2015 FDCCI deadline, almost all agencies still need to complete their inventories and consolidation plans and continue to identify additional targets for closure. Because closing facilities is a significant driver in realizing consolidation savings, the time required to realize planned cost savings will likely extend beyond the current 2015 time frame. With at least one agency not planning on realizing savings until after 2015 and other agencies having not yet reported on planned savings, there is an increased likelihood that agencies will either need more time to meet the overall FDCCI savings goal or that there are additional savings to be realized in years beyond 2015. Until OMB tracks cost savings data, the agency will be limited in its ability to determine whether or not FDCCI is on course toward achieving planned performance goals. Additionally, extending the horizon for realizing planned cost savings could provide OMB and FDCCI stakeholders with input and information on the benefits of consolidation beyond OMB's initial goal.

---

[25]GAO-12-742.

[26]One agency—the Department of Defense—estimated about $2.2 billion in cost savings by the end of 2015, which accounts for about 92 percent of the total anticipated cost savings. However, the department's consolidation plan noted that their cost savings estimates do not account for any up-front costs associated with the consolidation effort.

Oversight of FDCCI is Not Being Performed in All Key Areas

We have previously reported that oversight and governance of major IT initiatives help to ensure that the initiatives meet their objectives and performance goals.[27] When an initiative is governed by multiple entities, the roles and responsibilities of those entities should be clearly defined and documented, including the responsibilities for coordination among those entities. We have further reported,[28] and OMB requires,[29] that an executive-level body be responsible for overseeing major IT initiatives. Among other things, we have reported that this body should have documented policies and procedures for management oversight of the initiative, regularly track progress against established performance goals, and take corrective actions as needed.

Oversight and governance of FDCCI is the responsibility of several organizations—the Task Force, the GSA FDCCI Program Management Office, and OMB. Roles and responsibilities for these organizations are documented in the Task Force charter and OMB memoranda, while others are described in OMB's January 2013 quarterly report to Congress or have been communicated by agency officials. See table 1 for a listing of the FDCCI oversight and governance entities and their key responsibilities.

---

[27] GAO, *USDA Systems Modernization: Management and Oversight Improvements Are Needed*, GAO-11-586 (Washington, D.C.: July 20, 2011); *United States Coast Guard: Improvements Needed in Management and Oversight of Rescue System Acquisition*, GAO-06-623 (Washington, D.C.: May 31, 2006); and *Information Technology Investment Management: A Framework for Assessing and Improving Process Maturity*, GAO-04-394G (Washington, D.C.: March 2004).

[28] GAO-11-586, GAO-06-623, and GAO-04-394G.

[29] OMB, *Capital Programming Guide, V. 3.0, Supplement to OMB Circular A-11: Planning, Budgeting, and Acquisition of Capital Assets*, (Washington, D.C.: July 2012). The Capital Programming Guide is intended to help agencies effectively plan, procure, and use capital assets.

Table 1: Key Responsibilities of FDCCI Oversight and Governance Organizations

Organization	FDCCI responsibilities
Task Force	• Hold monthly meetings to facilitate communication between agencies • Communicate and coordinate agency best practices • Identify policy and implementation issues that could negatively impact agencies' abilities to meet their FDCCI goals • Assist agencies with development of their consolidation plans • **Develop and manage the data center total cost of ownership model (GSA)**[a] • **Develop an electronic governmentwide marketplace for data center availability (GSA and OMB)** • Oversee the agency consolidation plan peer review process
GSA FDCCI Program Management Office	• Collect agencies' consolidation inventories and plans annually (inventories in June; plans in September) • Under OMB direction, collect and disseminate data related to the FDCCI closure updates • Provide ad-hoc and quarterly reports to OMB regarding agencies' reported consolidation updates • Maintain and update FDCCI-related online portals, such as http://CIO.gov and http://data.gov (i.e., the consolidation progress dashboard) • **Develop and manage the data center total cost of ownership model (Task Force)** • Provide agency support including technical assistance on the total cost of ownership model and clarifying inventory and plan requirements • **Develop an electronic governmentwide marketplace for data center availability (Task Force and OMB)** • Conduct analysis of FDCCI inventories and plans, including reviewing agencies' submissions for errors
OMB	• Establish and manage FDCCI • Provide Federal CIO policy and guidance to the initiative • Launch an electronic public dashboard to track consolidation progress • **Develop an electronic governmentwide marketplace for data center availability (Task Force and GSA)** • Approve federal agency consolidation plans • Report quarterly to the Senate and House Committees on Appropriations identifying the savings achieved by OMB's governmentwide IT reform effort, which includes FDCCI • Provide executive-level oversight of FDCCI

Source: GAO analysis of OMB data and agency interviews.

Note: Bold text indicates a responsibility that is shared with the entities indicated in parentheses.

[a]This model is intended to provide agencies with a comprehensive tool to help inform decision making, model paths, and the development of cost savings figures and funding needs.

The Task Force, the GSA FDCCI Program Management Office, and OMB have performed a wide range of FDCCI responsibilities. For example, the Task Force holds monthly meetings to, among other things, communicate and coordinate consolidation best practices and to identify policy and implementation issues that could negatively impact the ability of agencies to meet their goals. Further, the Task Force has assisted agencies with the development of their consolidation plans by discussing lessons learned during its monthly meetings and disseminating new OMB guidance. GSA has collected responses to OMB-mandated document deliveries, including agencies' consolidation inventories and plans, on an annual basis. In addition, GSA has collected data related to FDCCI data center closure updates, disseminated the information publically on the consolidation progress dashboard on http://data.gov, and provided ad hoc and quarterly updates to OMB regarding these data. Lastly, as the executive-level body, OMB issued FDCCI policies and guidance in a series of memoranda that, among other things, required agencies to provide an updated data center asset inventory at the end of every third quarter and an updated consolidation plan at the end of every fourth quarter. In addition, OMB launched a publically available electronic dashboard to track and report on agencies' consolidation progress.

However, oversight of FDCCI is not being performed in other key areas. For example,

- The Task Force has not provided oversight of the agency consolidation peer review process. According to officials, the purpose of the peer review process is for agencies to get feedback on their consolidation plans and potential improvement suggestions from a partner agency with a data center environment of similar size and complexity. While the Task Force documented the agency pairings for 2011 and 2012 reviews, it did not provide agencies with guidance for executing their peer reviews, including information regarding the specific aspects of agency plans to be reviewed and the process for providing feedback. As a result, the peer review process did not ensure that significant weaknesses in agencies' plans were being identified. As previously mentioned, in July 2012, we reported[30] that all of the agencies' plans were incomplete except for one. In addition, we noted that three agencies had submitted their June 2011 inventory updates, a required component of consolidation documentation, in an incorrect format—an outdated template.

- The GSA FDCCI Program Management Office has not executed its responsibilities related to analyzing agencies' inventories and plans and reviewing these documents for errors. In July 2012, we reported on agencies' progress toward completing their inventories and plans

---

[30]GAO-12-742.

and found that only three agencies had submitted a complete inventory and only one agency had submitted a complete plan, and that most agencies did not fully report cost savings information and eight agencies did not include any cost savings information.[31] The lack of cost savings information is particularly important because, as previously noted, initiativewide cost savings have not been determined—a shortcoming that could potentially be addressed if agencies had submitted complete plans that addressed cost savings realized, as required.

- Although OMB is the approval authority of agencies' consolidation plans, it has not approved agencies' submissions on the basis of their completeness. In an October 2010 memorandum, OMB stated that its approval of agencies' consolidation plans was in progress and would be completed by December 2010. However, OMB did not issue a subsequent memorandum indicating that it had approved agencies' plans, or an updated time frame for completing its review. This is important because, in July 2011 and July 2012, we reported that agencies' consolidation plans had significant weaknesses and that nearly all were incomplete.[32]

- OMB has not reported on agencies' progress against its key performance goal of achieving $3 billion in cost savings by the end of 2015. Although the 2012 Consolidated Appropriations Act included a provision directing OMB to submit quarterly progress reports to the Senate and House Appropriations Committees that identify savings achieved through governmentwide IT reform efforts,[33] OMB has not yet reported on cost savings realized for FDCCI. Instead, the agency's quarterly reports had only described planned FDCCI-related savings and stated that future reports will identify savings realized. As of the January 2013 report, no such savings have been reported.

These weaknesses in oversight are due, in part, to OMB not ensuring that assigned responsibilities are being executed. Improved oversight could better position OMB to assess progress against its cost savings goal and minimize agencies' risk of not realizing anticipated cost savings.

---

[31]GAO-12-742.

[32]GAO-11-565 and GAO-12-742.

[33]Consolidated Appropriations Act, 2012, Pub. L. No. 112-74, div. C, title II, 125 Stat. 786, 896 (2011).

Recent Integration with PortfolioStat Changes FDCCI, but Reporting Requirements and Goals Are Not Fully Defined

OMB's recent integration of FDCCI and PortfolioStat made significant changes to data center consolidation oversight and reporting requirements. According to OMB's March 2013 memorandum,[34] to more effectively measure the efficiency of an agency's data center assets, agency progress will no longer be measured solely by closures. Instead, agencies will also be measured by the extent to which their data centers are optimized for total cost of ownership by incorporating metrics for energy, facility, labor, and storage, among other things. In addition, OMB stated that the Task Force will categorize agencies' data center populations into two categories—core and non-core data centers—for which the memorandum does not provide specific definitions. Additionally, as previously discussed, agencies are no longer required to submit the data center consolidation plans previously required under FDCCI. Rather, agencies are to submit information to OMB via three primary means—an information resources management strategic plan, an enterprise road map, and an integrated data collection channel. Using these tools, an agency is to report on, among other things, its approach to optimizing its data centers; the state of its data center population, including the number of core and non-core data centers; the agency's progress on closures; and the extent to which an agency's data centers are optimized for total cost of ownership.

However, OMB's memorandum does not fully address the revised goals and reporting requirements of the combined initiative. Specifically, OMB stated that its new goal is to close 40 percent of non-core data centers but, as previously mentioned, the definitions for core and non-core data center were not provided. Therefore, the total number of data centers to be closed under OMB's revised goal cannot be determined. In addition, although OMB has indicated which performance measures it plans to use going forward, such as those related to data center energy and labor, it has not documented the specific metrics for agencies to report against. The memorandum indicates that these will be developed by the Task Force, but does not provide a time frame for when this will be completed. Lastly, although OMB has previously stated that PortfolioStat is expected to result in savings of approximately $2.5 billion through 2015, its memorandum does not establish a new cost savings goal for FDCCI, nor does it refer to the previous goal of saving $3 billion. Instead, OMB states that all cost savings goals previously associated with FDCCI will be

---

[34]OMB Memorandum, M-13-09.

integrated into broader agency efforts to reshape their IT portfolios, but does not provide a revised savings estimate. The lack of a new cost savings goal will further limit OMB's ability to determine whether or not the new combined initiative is on course toward achieving its planned objectives.

In addition, several important oversight responsibilities related to data center consolidation have not been addressed. For example, with the elimination of the requirement to submit separate data center consolidation plans under the new combined initiative, the memorandum does not discuss whether either the Task Force or the GSA Program Management Office will continue to be used in their same oversight roles for review of agencies' documentation. In addition, while the memorandum discusses OMB's responsibility for reviewing agencies' draft strategic plans, it does not discuss the responsibility for approving them. In the absence of defined oversight assignments and responsibilities, it cannot be determined how OMB will have assurance that agencies' plans meet the revised program requirements and, moving forward, whether these plans support the goals of the combined initiative.

Implementation of Recommendations Could Help Ensure Improvements in Oversight

In our report being released today, we are making recommendations to better ensure that FDCCI achieves expected cost savings and to improve executive-level oversight of the initiative. Specifically, we are recommending that the Director of OMB direct the Federal CIO to

- track and annually report on key data center consolidation performance measures, such as the size of data centers being closed and cost savings to date;

- extend the time frame for achieving cost savings related to data center consolidation beyond the current 2015 horizon, to allow time to meet the initiative's planned cost savings goal; and

- establish a mechanism to ensure that the established responsibilities of designated data center consolidation oversight organizations are fully executed, including responsibility for the documentation and oversight of the peer review process, the review of agencies' updated consolidation inventories and plans, and approval of updated consolidation plans.

The Federal CIO stated that the agency concurred with the first and third recommendation. Regarding the second recommendation, OMB neither agreed nor disagreed. However, the Federal CIO stated that, as the

FDCCI and PortfolioStat initiatives proceed and continue to generate savings, OMB will consider whether updates to the current time frame are appropriate.

In summary, after more than 3 years into FDCCI, agencies have made progress in their efforts to close data centers. However, many key aspects of the integration of FDCCI and PortfolioStat, including new data center consolidation and cost savings goals, have not yet been defined. Further compounding this lack of clarity, total cost savings to date from data center consolidation efforts have not been determined, creating uncertainty as to whether OMB will be able to meet its original cost savings goal of $3 billion by the end of 2015. In the absence of tracking and reporting on cost savings and additional time for agencies to achieve planned savings, OMB will be challenged in ensuring that the initiative, under this new direction, is meeting its established objectives.

Recognizing the importance of effective oversight of major IT initiatives, OMB directed that three oversight organizations—the Task Force, the GSA FDCCI Program Management Office, and OMB—be responsible for federal data center consolidation oversight activities. These organizations have performed a wide range of FDCCI responsibilities, including facilitating collaboration among agencies and developing tools to assist agencies in their consolidation efforts. However, other key oversight activities have not been performed. Most notably, the lack of formal guidance for consolidation plan peer review and approval increases the risk that missing elements will continue to go undetected and that agencies' efforts will not fully support OMB's goals. Further, OMB's March 2013 memorandum does not address whether the Task Force and GSA's Program Management Office will continue their oversight roles, which does not help to mitigate this risk. Finally, while OMB has put in place initiatives to track consolidation progress, consolidation inventories and plans are not being reviewed for errors and cost savings are not being tracked or reported. The collective importance of these activities to federal data center consolidation success reinforces the need for oversight responsibilities to be fulfilled in accordance with established requirements.

Chairman Mica, Ranking Member Connolly, and Members of the Subcommittee, this completes my prepared statement. I would be pleased to respond to any questions that you may have at this time.

GAO Contact and Staff Acknowledgments

If you or your staffs have any questions about this testimony, please contact me at (202) 512-9286 or at pownerd@gao.gov. Individuals who made key contributions to this testimony are Dave Hinchman (Assistant Director), Justin Booth, Nancy Glover, and Jonathan Ticehurst.

This is a work of the U.S. government and is not subject to copyright protection in the United States. The published product may be reproduced and distributed in its entirety without further permission from GAO. However, because this work may contain copyrighted images or other material, permission from the copyright holder may be necessary if you wish to reproduce this material separately.

GAO's Mission	The Government Accountability Office, the audit, evaluation, and investigative arm of Congress, exists to support Congress in meeting its constitutional responsibilities and to help improve the performance and accountability of the federal government for the American people. GAO examines the use of public funds; evaluates federal programs and policies; and provides analyses, recommendations, and other assistance to help Congress make informed oversight, policy, and funding decisions. GAO's commitment to good government is reflected in its core values of accountability, integrity, and reliability.
Obtaining Copies of GAO Reports and Testimony	The fastest and easiest way to obtain copies of GAO documents at no cost is through GAO's website (http://www.gao.gov). Each weekday afternoon, GAO posts on its website newly released reports, testimony, and correspondence. To have GAO e-mail you a list of newly posted products, go to http://www.gao.gov and select "E-mail Updates."
Order by Phone	The price of each GAO publication reflects GAO's actual cost of production and distribution and depends on the number of pages in the publication and whether the publication is printed in color or black and white. Pricing and ordering information is posted on GAO's website, http://www.gao.gov/ordering.htm. Place orders by calling (202) 512-6000, toll free (866) 801-7077, or TDD (202) 512-2537. Orders may be paid for using American Express, Discover Card, MasterCard, Visa, check, or money order. Call for additional information.
Connect with GAO	Connect with GAO on Facebook, Flickr, Twitter, and YouTube. Subscribe to our RSS Feeds or E-mail Updates. Listen to our Podcasts. Visit GAO on the web at www.gao.gov.
To Report Fraud, Waste, and Abuse in Federal Programs	Contact: Website: http://www.gao.gov/fraudnet/fraudnet.htm E-mail: fraudnet@gao.gov Automated answering system: (800) 424-5454 or (202) 512-7470
Congressional Relations	Katherine Siggerud, Managing Director, siggerudk@gao.gov, (202) 512-4400, U.S. Government Accountability Office, 441 G Street NW, Room 7125, Washington, DC 20548
Public Affairs	Chuck Young, Managing Director, youngc1@gao.gov, (202) 512-4800 U.S. Government Accountability Office, 441 G Street NW, Room 7149 Washington, DC 20548

Please Print on Recycled Paper.

www.ingramcontent.com/pod-product-compliance
Lightning Source LLC
Chambersburg PA
CBHW081821170526
45167CB00008B/3495